Machine Learning in Healthcare

Vaibhav Rupapara

Table of Contents

ABOUT THE BOOK

Have you ever come across the subject of Machine Learning in Health Care services? Well, this very book is designed to look intently and intensely into the nitty-gritty of how Machine Learning, a form of artificial intelligence, is deployed meaningfully in the health care sector. The author intends to answer the questions about the meaning of Machine Learning and how it is applied in the health care systems, and the various benefits and drawbacks associated with this product of the 21st century. In addition, the great potentials of this machine learning in the health care system, especially what it holds for the future, is also meticulously considered. Happy Reading!

INTRODUCTION

It is total with presumably that the coming of digitalization caused a type of interruption in each industry, including the medical care area. The capacity to catch, share and convey information is turning into the highest need. AI, extensive knowledge, and artificial brainpower (simulated intelligence) can address the soar quantum of information's various difficulties. AI has the capacity to help medical services suppliers satisfy developing clinical needs, further develop tasks and lower costs. The wording "AI" was imagined and characterized as "... counterfeit age of information for a fact." The preliminary examinations have been performed with games, i.e., with the round of checkers. Be that as it may, Today, AI (ML) is the quickest developing specialized field, at the convergence of informatics and insights, firmly associated with information science and information disclosure, and well-being is among the best difficulties going up against people. Especially, probabilistic AI is beneficial for well-being informatics, where most issues include managing vulnerability. The hypothetical reason for the probabilistic AI, for example, was laid by Thomas Bayes (1701–1761). Probabilistic induction boundlessly affected artificial brainpower and authentic learning, and the converse likelihood permits construing questions, gaining information, and making forecasts about phenomena.

It will give much joy to acknowledge that despite the delayed improvement in ML has been engineered both by the enhancement of rejuvenated learning measurements and studies and by the ongoing blast of data and, simultaneously, minimal expense calculation. The reception of information escalated AI calculations can be found in all application spaces of well-being informatics and is especially helpful for mind informatics, going from essential exploration to comprehend insight to a broad scope of explicit cerebrum informatics research. The utilization of AI strategies in biomedicine and well-being can, for example, lead to more proof-based dynamics and assisting with going toward customized medication. Outstandingly, as per Tom Mitchell, a logical field is best characterized by the inquiries it contemplates: Subsequently, AI looks to respond to the question consequently;

"How might we assemble calculations that naturally work on through experience, and what are the key laws that administer all learning measures?"

Simply have it at the rear of your brain if you are in the clinical field. Machine Learning development can help medical care specialists distinguish and treat illness more productively and with more accuracy and customized care. An assessment of Machine Learning in medical services uncovers how innovation advancement can prompt more robust, comprehensive consideration systems that could work on tolerant results and work on their experience on an exceptionally all-encompassing level.

CHAPTER ONE

MACHINE LEARNING

The machine is the most well-known type of Artificial Learning. It cycles and discovers designs in huge informational collections to empower dynamics. Machine Learning applications comprise calculations: an assortment of directions for playing out a particular arrangement of errands. The calculations are intended to gain from the information autonomously, without human mediation. Over the long run, Machine Learning calculations further develop their forecast exactness without requiring programming. A profound plunge into what AI is uncovered three basic parts of calculations: portrayal, assessment and streamlining. First, portrayal implies that information should be arranged in a structure and language that a PC can deal with. This segment makes way for the following part, assessment, to decide if the information groupings are helpful. Then, at that point, as a component of the improvement interaction, the calculation tracks down the best model for the best and precise yields.

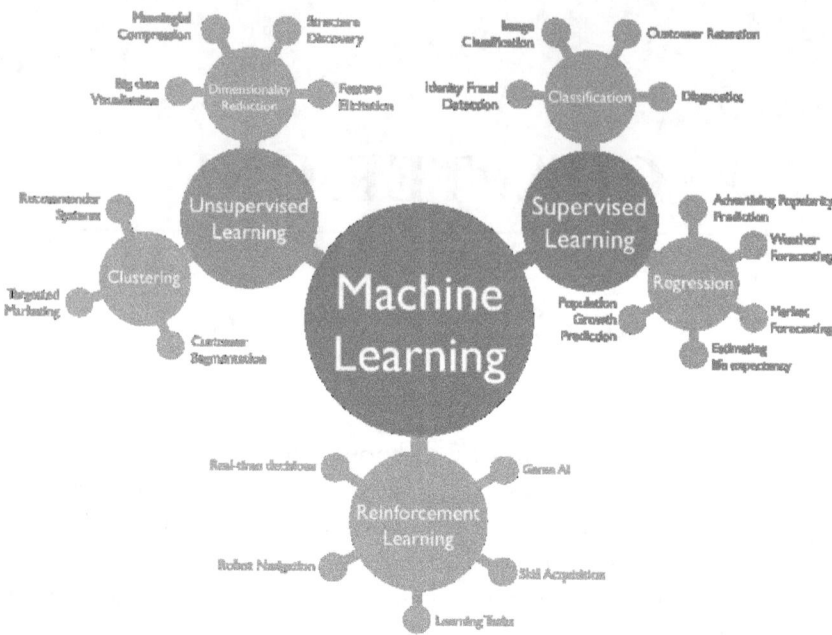

Source: towardsdatascience.com

Machine Learning is an integral part of the emerging branch of data science. Optimizing measurable techniques, calculations are ready to effect expectations, showing core experiences inside information mining projects. This avalanche of knowledge, therefore, pushes dynamic inside networks and corporations, favorably affecting core enhancement measurements and considering the level of more data outputs to enlarge and improve the commercial priority for resource explorers with increment, anticipating who to assist the ID of the most significant enterprise queries within the scope of the available database to react effectively therein.

Be that as it may, a vital qualification must be noted to keep away from a circumstance where you will befuddle comparative however totally different terms. Deep learning and machine learning can be optimized conversely. Machine Learning, Deep learning, and neural networks are majorly sub-sets of the digitalized thinking process. Be that as it may, deep learning is a subset of Artificial Intelligence, and neural networks, on the other hand, finds many expressions in deep learning. The medium through which this deep learning (DL) and Machine Learning (ML) are divergent appears to be in their respective calculation processes. DL houses a core aspect of the ingredient piece of the interaction, dispensing with a big fraction of the traditional human intercession needed and enabling the usage of sizable informational indices. You will not be entirely out of point if you look at Deep learning as "versatile AI," just like Lex Fridman notes in his MIT address (00:30) (connect lives outside IBM). Traditional, or "non-profound," ML is dependent on how humans interact within their environment to learn. Experts in human behavior determine the placement

of elements to understand the distinctions between data inputs, for the greater part needing a lot of well-coordinated information to learn.

Neural networks, or Artificial neural systems (ANS), are layers comprising a data layer, at least one hidden layer, and a productive layer. Every hub, or artificial neuron, interconnects with another and has a similar weight and limit. Upon the incidental reality that the productivity of any personal combination is above the pre-conceived edge estimation, that hub is actuated, delivering data to the subsequent layer of the organization. Additionally, no data is offered to the following layer of the networks. The "Deep" in Deep learning is ordinarily referring to the profundity of layers in a neural networking apparatus. A neural network that consists of plenteous layers, which may be holistic of the sources of info and the outcome, can be seen as a deep learning calculation or deep neural networking.

Types of Machine Learning

Supervised Learning

Unsupervised Learning

Reinforcement Learning

Source: simplilearn.com

Types of Machine Learning

You should know that Machine Learning, which is an aspect of artificial intelligence is clearly in four major types, and they have been reproduced thus:

a) Supervised machine learning

Supervised learning is characterized by its utilization of marked datasets to prepare calculations that arrange information or anticipate results precisely. As information is taken care of into the model, it changes its loads until the model has been fitted. This happens as a component of the cross-approval interaction to guarantee that the model keeps away from overfitting or underfitting. Managed learning assists associations with addressing an assortment of certifiable issues at scale, for example, grouping spam in a different organizer from your inbox. A few strategies utilized in administered learning incorporate neural organizations, guileless Bayes, straight relapse, calculated relapse, irregular woodland, support vector machine (SVM), and that's only the tip of the iceberg. Supervised learning can be additionally isolated into two sorts of issues when information mining: **Classification and Regression:** Classification issues utilize a calculation to precisely allocate test information into explicit classifications, like isolating apples from oranges. Or then again directed learning calculations can be utilized to characterize spam in a different organizer from your inbox. Direct classifiers, support vector machines, choice trees, and irregular backwoods are altogether normal sorts of characterization calculations. Regression is another administered learning technique that utilizes a calculation to comprehend the connection among reliant and free factors. Regression models are useful for foreseeing mathematical qualities dependent on various information focuses, for example, deals income projections for a given business.

b) Unsupervised machine learning:

This is also known as unsupervised learning. It utilizes ML Algorithms to examine and categorize unlabeled sets of data. Hidden patterns are discovered using this Algorithm. After the discovery, they are categorized with human supervision or interference. Machine learning is used to explore data and carry out analysis, pattern detection, image recognition, client categorization because not its similarities and disparities detection feature. This makes it the ideal tool to use.; Principal Component Analysis (PCA) and Singular Value Decomposition (SVD) are two ordinary philosophies for this Machine Learning. Various estimations used in independent learning join neural associations, k-suggests bundling, probabilistic gathering systems, and that is just a glimpse of something larger. Unsupervised learning models are utilized for three primary undertakings: bunching, affiliation, and dimensionality decrease: Bunching is an information-digging procedure for gathering unlabeled information dependent on their likenesses or contrasts. For instance, K-implies bunching calculations relegate comparable information focuses into gatherings, where the K worth addresses the size of the gathering and granularity. This procedure is useful for market division, picture pressure, and so on. Affiliation is another solo learning technique that utilizes various principles to discover connections between factors in each dataset. These strategies are regularly utilized for market crate investigation and proposal motors, as per "Clients Who Purchased This Thing Additionally Purchased" suggestions. Dimensionality decrease is a learning procedure utilized when the quantity of elements (or measurements) in each dataset is excessively high. Thus, it lessens the number of information contributions to a reasonable size while protecting information honesty. This procedure is frequently utilized in the preprocessing information stage, for example, when auto encoders eliminate commotion from visual information to develop picture quality further.

c) Semi-supervised learning:

Semi-supervised learning is an approach to manage ML that combines a discreet amount of checked data with a great deal of unlabeled data during getting ready. Semi-coordinated learning falls between solo learning (with no stamped getting ready data) and regulated learning (with just named planning data). It is an exceptional illustration that weak Unlabeled data, when used identified with a humble amount of named data, can make broad improvement in learning accuracy. However, the getting of checked data for a learning issue routinely requires a skilled human subject matter expert (for instance, to decipher a sound piece) or a genuine assessment (for instance, choosing the 3D development of a protein or choosing if there is oil at a particular region). This way, the cost related to the checking collaboration may convey hugely, totally named getting ready sets infeasible. However, the acquirement of unlabeled data is modest. In such conditions, semi-oversaw learning can be of mind-blowing useful worth. Semi-oversaw learning is also of theoretical premium in artificial intelligence and as a model for human learning.

Human reactions to formal semi-directed learning issues have yielded changing decisions about the level of impact of the unlabeled information. More normal learning issues may likewise be seen as examples of semi-regulated learning. A lot of human idea learning includes a modest quantity of direct guidance (for example, parental naming of articles during youth) joined with a lot of unlabeled experience (for example, perception of items without naming or checking them, or without criticism). Human newborn children are touchy to the construction of unlabeled normal classifications like pictures of canines and felines or male and female appearances. Babies and kids consider unlabeled models. However, the inspecting cycle from which marked models emerge.

d) Reinforcement learning

Reinforcement learning (RL) was examined by Turing and is to date the most contemplated approach in AI. The hypothesis behind Reinforcement Learning (otherwise known as "Support Learning") is established in neuropsychological issues on the conduct of how specialists might upgrade their control of an unpredictable climate. Subsequently, Support Learning is a part of Machine Learning that is worried about utilizing experience acquired through associating with the world and evaluative criticism to work on the capacity of a framework to create conduct choices. This has been known as the artificial consciousness issue in a microcosm since learning specialists should perform well and accomplish their objectives independently. Driven by the expanding accessibility of rich information, Reinforcement learning has accomplished extraordinary outcomes, remembering advancements for essential ML-significant regions, like speculation, arranging, investigation, and exact procedure, prompting better relevance to genuine issues.

Support Taking in is not the same as managed to realize, where taking in occurs from models given by an outside (human) boss. This is a significant sort of learning; nonetheless, it alone isn't adequate for gaining from collaboration. In intelligent issues, it is normally unfeasible to get instances of wanted conduct that are both right and delegate for every one of the circumstances in which the specialist needs to act. In obscure regions (which we need to investigate), where one would anticipate that learning should be gainful—a specialist should have the option to gain from its own insight. Regardless of all restrictions, Support Learning is the principal field to truly resolve the computational issues that emerge when gaining from connection with a climate to accomplish long haul objectives, since it utilizes a proper structure characterizing the association between a learning specialist and its current circumstance as far as states, activities, and prizes. This structure is expected to be a basic method of addressing fundamental elements of general Man-made reasoning issues and components, including a feeling of circumstances and logical results, a feeling of vulnerability and non-determinism, and the presence of unequivocal objectives. In the average Support Learning-model, a specialist is associated with its current circumstance employing discernment and activity. In each progression of collaboration, the specialist gets as information I, some sign of the present status s, of the climate. Afterward, the specialist picks an activity a, which changes the condition of the climate. The worth of this state change is conveyed to the specialist through a scalar support signal r. The specialist's conduct B should now pick activities that will, in general, expand the since quite a while ago run number of upsides of the Reinforcement Learning signal. It can figure out how to do this over the long haul by orderly experimentation, directed by a wide assortment of calculations.

Understanding the application of Machine Learning

Any assignment that can be finished with information characterized example or set of rules can be mechanized with Machine Learning. This permits organizations to change measures that were beforehand an option exclusively for people to perform—think reacting to client support calls, accounting, and auditing resumes. From mechanizing dreary manual information sections to more perplexing use cases like protection hazard appraisals or extortion location, Machine Learning has numerous applications, including customer confronting capacities like client support, item proposals, and inward applications inside associations to help accelerate measures lessen manual responsibilities. A significant piece of what makes AI so important is its capacity to identify what the natural eye misses. AI models can get perplexing examples that would have been neglected during a human examination. Much gratitude to intellectual innovation like normal language handling, machine vision, and profound learning. Machine learning opens human specialists to zero in on assignments like item advancement and consummating administration quality and effectiveness. You may be acceptable at filtering through a fatal yet well-articulated accounting platform. However, because of ML and digitalized rationalization, measurements can look at a lot bigger arrangements of information and understand designs significantly and more rapidly.

Most information researchers are acquainted with how R and Python programming dialects are utilized for Machine Learning. There are a lot of other dialectic potentials also, contingent upon the kind of model or undertaking needs. Machine learning and Man-made brainpower instruments regularly program libraries, tool stash, or suites that guide executing assignments. Be that as it may, due to its inescapable help and the large number of libraries to browse, Python is viewed as the most well-known programming language for Machine Learning. Truth be told, as per GitHub, Python is number one on the rundown of the top Machine Learning dialects on their site. Python is regularly utilized for information mining and information examination and supports the execution of a wide scope of AI models and calculations.

Upheld calculations in Python incorporate order, relapse, grouping, and dimensionality decrease. However, Python is the main language in AI, and a few others are exceptionally well known. Since some Machine Learning applications use models written in various dialects, apparatuses like AI activities (MLOps) can be especially useful. Moreover, AI can offer some benefit to customers just as to endeavors. For example, a venture can acquire knowledge into its serious scene and client steadfastness and figure deals or interest continuously with Machine Learning. Note also that Machine learning operations (MLOps) are the discipline of Artificial Intelligence model delivery.

Some real-world cases where machine learning is used include:

- **Speed Recognition**: Artificial intelligence through machine learning has been deployed for ASR (Automatic Speech Recognition) purposes. It could be text to speech. The AI uses Natural Language Processing to analyze convert speeches to writings. Many smartphone devices have incorporated the speech recognition system into their devices. An example is Apple's software care Siri which also does more than transcribing speech.

- **Customer service**: Many industries with websites use chatbots to answer frequently asked questions. This has replaced the need for human agents. Areas such as marketing, making the simplest suggestions from users, renewed customers engagement across social media platforms. For instance, chatbots on marketing sites with virtual agents and even social media apps like Facebook usually use virtual or voice assistants.

- **Computer's vision**: Data can be mined or extracted from videos, virtual inputs like Digital images, and computer systems using this Artificial intelligence technology. This is different from image recognition because it provides outputs, results, and suggestions based on the mined data. Convolutional neural networks power computer vision, and it has been used by some social media apps such as Instagram and Facebook to tag within photos. Various industries have also applied this AI technology.

- **Recommendation Engines**: Data can be mined using a history of consumption behavior. The AI Algorithms detects data pattern that can be utilized in marketing for more effective cross-selling techniques. Through data gathering, the AI finds patterns and sequences. It then uses this pattern for targeted add-on suggestions during the process of payments for online retailers.

- **Automated Stock trading**: AI-driven trading platforms make many transactions per day with human agents interfering. They are structured to improve stock portfolios using collected data.

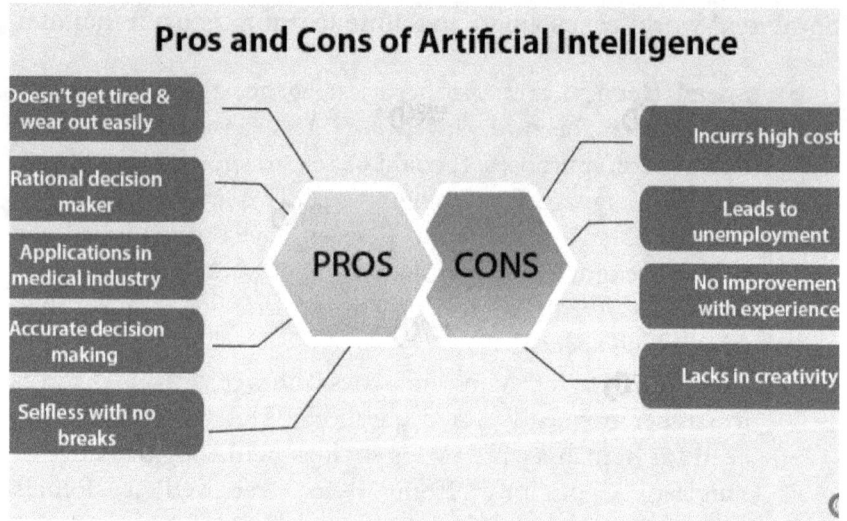

Pros and Cons of Artificial Intelligence

Doesn't get tired & wear out easily

Rational decision maker

Applications in medical industry

Accurate decision making

Selfless with no breaks

PROS   CONS

Incurrs high cost

Leads to unemployment

No improvement with experience

Lacks in creativity

Source: data-flair.training

Drawbacks Associated with Machine Learning and Artificial Intelligence application

a) **Information Obtaining**

Machine Learning requires gigantic informational indexes to prepare on, and this ought to be comprehensive/fair and of good quality. There can likewise be times where they should trust that new information will be produced.

b) **Time and Assets**

Machine Learning needs sufficient opportunity to allow the calculations to learn and foster to satisfy their motivation with a lot of exactness and importance. It additionally needs huge assets to work. This can mean extra necessities of PC power for you.

c) **Understanding of Results**

Another significant test is the capacity to decipher the results produced by the calculations precisely. Therefore, you should likewise cautiously pick the calculations for your motivation.

d) **High mistake Vulnerability**

Machine Learning stands alone yet incredibly defenseless to blunders. Suppose you train an accountant with some collections of data little not to be comprehensive. You are left with a lopsided prediction coming from a skewed arranging set. Do know that it ignites

18

superfluous notices being shown to customers. Due to the presence of Machine Learning, such bungles can precipitate a chain of errors holding sway without been detected for a long time.

Note that these bungles are a typical issue that is capable ordinarily. Since when these missteps occur, it isn't difficult to discover the fundamental hotspot for which the issue has been made and to discover that specific issue and correcting it takes a more extended time.

CHAPTER TWO

MACHINE LEARNING APPLICATION IN HEALTH CARE

Methods of applying Machine Learning in Health Care services

Medical care is a significant industry that offers esteem-based consideration to many individuals while simultaneously becoming top income workers for some nations. Today, the medical care industry in the US alone acquires an income of $1.668 trillion. The US additionally spends more on medical services per capita when contrasted with most other created or non-industrial countries. Quality, worth, and results are three popular expressions that consistently go with medical services and guarantee a great deal. Today, medical care-trained professionals and partners throughout the planet are searching for imaginative approaches to follow through on this guarantee. Innovation empowered shrewd medical services is at this point not a trip of extravagant, as Web associated clinical gadgets are holding the well-being framework as far as we might be concerned together from self-destructing under the populace trouble.

Source: techiexpert.com

From assuming a basic part in persistent consideration, charging, and clinical records, today's innovation permits medical care experts to foster substitute staffing models, IP capitalization, give brilliant medical care, and lessening authoritative and supply costs. Machine Learning in medical services is one such region that sees continuous acknowledgment in the medical services industry. Google, as of late, fostered an AI calculation to distinguish harmful tumors in mammograms, and analysts in Stanford College are utilizing profound figuring out how to recognize skin malignant growth. Machine Learning (ML) is now helping in different circumstances in medical services. Machine Language in medical services assists with investigating many various information focuses and propose results, give opportune danger scores, exact asset portion, and has numerous different applications.

Consequently, there are some methodologies of applying Machine Learning in Health Care services, and they are:

a) Detecting Ailments and Diagnosis

A major aspect of Machine learning applications in clinical consideration is the ID and finish of infections and illnesses, which are seen as hard to dissect. This can fuse anything from tumors that are difficult to find during the fundamental stages to other innate ailments.

IBM Watson Genomics is an incredible portrayal of how joining mental preparation with genome-based tumor numbering can help make a fast finding. Berg, the biopharma beast, is using Man-made intellectual competence to cultivate therapeutic drugs in locales like oncology. P1vital's Predict (Expecting Response to Debilitation Treatment) hopes to cultivate a monetarily attainable way to deal with dissecting and give treatment in routine clinical conditions.

b) Medical Imaging and Assembling

One of the fundamental medical purposes of Machine learning lies in starting stage drug disclosure measures. This moreover joins Innovative work developments, for instance, forefront sequencing and exactness drugs, which can help find elective ways to treat multifactorial contaminations. As of now, the artificial intelligence strategies incorporate performance acknowledging, which can recognize plans in data without giving any conjectures. Adventure Hanover, made by Microsoft, is using computer-based intelligence-based advancements for different drives, including making artificial insight-based development for infection therapy and redoing drug mix for EML (Extraordinary Myeloid Leukemia).

c) Clinical Imaging Finding

Machine Learning and Profound learning are both liable for the progression of development called PC Vision. This has found attirmation in the Inner Eye drive made by Microsoft, which manages picture decisive mechanical assemblies for picture analysis. As computer-based intelligence ends up being very porous and as they fill in their illustrative limit, the desire to see more informational resources from changed clinical imagery becomes a piece of this artificial knowledge-driven insightful cycle.

d) Altered Drug

Modified medications can not only be more practical by coordinating with singular prosperity with perceptive assessment; however, then again is prepared for extra investigation and better infection assessment. As of now, specialists are confined to investigating a specific plan of discoveries or checking the patient ward's risk on his demonstrative history and open innate information. In any case, artificial intelligence in prescription is making unprecedented strides. IBM Watson Oncology is at the front line of this improvement by using patient clinical history to help make various treatment decisions. In the coming years, we will see more devices and biosensors with refined prosperity assessment limits hit the market, allowing more data to open for such cutting-edge ML-based clinical benefits propels.

e) AI-based Social Change

Social change is a huge piece of preventive medicine. Since the time of the increase of artificial intelligence in clinical benefits, interminable new organizations are jumping up in the fields of illness countering and conspicuous verification, patient treatment, etc. Somatix is a B2B2C-based data assessment association that has conveyed an ML-based application to see signals we make in our step-by-step lives, allowing us to fathom our unmindful lead and carry out crucial upgrades.

f) Clinical Starter and Investigation

Computer-based intelligence has a couple of potential applications in the field of clinical primers and investigation. First, as anybody in the pharma business would prompt you, clinical starters cost a lot of time and cash and can require quite a while to wrap up overall. Second, applying ML-based farsighted examination to perceive potential clinical starter contenders can help researchers draw a pool from a wide variety of data centers, for instance, past expert visits, online media, etc. artificial intelligence has moreover found usage in ensuring ceaseless checking and data access of the primer individuals, finding the best model size to be attempted, and using the power of electronic records to decrease data-based slip-ups.

g) Better Radiotherapy

Potentially the most sought-after usages of computer-based intelligence in clinical consideration is in the field of Radiology. Clinical picture examination has various discrete components which can arise at a particular depiction of time. For example, various wounds, danger foci, etc., can't be shown using complex conditions. Since ML-based computations acquire from the enormous number of different models open accessible, it becomes easier to investigate and find the variables. The most notable business of computer-based intelligence in clinical picture assessment is the portrayal of articles, for instance, wounds into arrangements like regular or uncommon, injury or non-sore, etc. Google's DeepMind Prosperity is helping examiners in University College London Hospitals cultivate estimations that can recognize the differentiation among strong and harmful tissue and further foster radiation treatment for the same.

h) Freely upheld Data Variety

Freely supporting is very well known in the clinical field nowadays, allowing investigators and specialists to get to a colossal proportion of information moved by people reliant upon their own consent. This live prosperity data has unbelievable outcomes in the way medicine will be seen down the line. For example, apple's Exploration Unit grants customers instinctive applications that apply ML-based facial affirmation to endeavor to treat Asperger's and Parkinson's contamination. IBM teamed up with Medtronic to decipher, total, and make available diabetes and insulin data persistently reliant upon the freely upheld information. With the movements being made in IoT, the clinical benefits industry is discovering new habits by which to use this data and tackle extraordinary to investigate cases and help improve examination and remedy.

i) Brilliant Electronic Wellbeing Recorder

AI extension, for example, the optical characters and report order can likewise be utilized to create with the keen electronic well-being record framework. The undertaking of this application is added to chip away at fostering a framework that can even sort the patient inquiries with the assistance of an email or even to change the manual record framework into a computerized apparatus framework. The essential target of this application is to work with a safe and effectively open framework.

The quick face pace development of electronic well-being records has likewise been advanced with the store of clinical information about the patients, which can likewise be utilized for the further developing medical care framework. It even lessens the information mistakes, for instance, copy information. To create and assemble the electronic well-being recorder framework, directed AI calculations like the help vector machine can be utilized as a classifier or the Counterfeit Neural Organization, which can be applied without any problem.

j) Medications

From the Sedation to the treatment from chest harm to step by step sedates, the use of artificial intelligence in clinical benefits is as of now being used in the overhauls of the prescriptions. The acclaimed IBM supercomputer Watson is in like manner working with associations like Pfizer to additionally foster medication exposure, likewise, from the immunological conditions and dangerous development. Google has similarly been in this game for such incalculable years and has also been seen to be more astounding with the opportunities for artificial intelligence to direct and chip away at the considerations around the meds. A part of the associations like Medtronic is also managing to utilize artificial intelligence to chip away at the prescriptions, nevertheless, significantly more on the individual scale. A segment of the redirecting musings regarding a precision drug, they are moreover working on with the patients to offer altered contribution on the ideal way to even more probable control and treat their diabetes. Redone medicine will be a fundamental strength of patients later.

Advantages of applying Machine Learning in Health Care services

There are lots of ways that how Machine learning in health care can be beneficial. Some top benefits have been identified as follows:

a) Interfacing Information

Information examination is a gigantic subject in medical services. As a public, we are progressively embracing computerized well-being, and as those increments, so does the measure of information and an assortment of information sources. This information flood is over-burdening medical services frameworks, and clinical experts are struggling to process the entirety of this crude information properly. Preparing information from various sources and giving a prescient examination is quite possibly the main advantage that artificial intelligence can bring to medical care. The medical care industry is continually tormented with the difficulty of having information from numerous various sources and figuring out how to unite and handle that information into valuable data for clinicians. Artificial intelligence can take divergent information from things like wearable gadgets, electronic well-being records, and lab testing and give specialists ideas for care dependent on that examination. This data can incorporate infection hazards, conclusion ideas, and example notices in an inexorably proficient and precise style.

b) Working on Persistent Consideration

Anticipating the opportunity of and recognizing infections with artificial intelligence is utilization that could really transform patients. Utilizing calculations with patient informational indexes and sources can help specialists and other clinical experts screen for sicknesses with an extremely undeniable degree of precision. However, the objective isn't to supplant clinical experts, for those experts to utilize simulated intelligence as clinical choice help and "another arrangement of eyes" to diminish the opportunity of blunders. For instance, utilizing computerized reasoning to deal with a patient's clinical records and lab records can assist with anticipating the odds of illnesses, including things like diabetes, cardiovascular infection, and so forth. Utilizing artificial intelligence to deal with this information can help medical services experts comprehend patient examples and see where potential patient necessities emerge. Artificial intelligence innovation can deal with a bigger number of information quicker than any human, making it an incredible commendation to any clinician's act of medication and an extremely proficient method of getting significant information.

26

c) Improved Patient Correspondence and Access

Advancing patient correspondence is vital when hoping to work on understanding consideration. Using innovations, for example, virtual nursing aides can assist with further developing correspondence with medical care suppliers and lessen readmissions or superfluous trauma center visits. Utilizing computer-based intelligence, virtual nursing associates can repeat nursing conduct and help with patient inquiries, checking, updates, and giving answers. When seeing patient access, particularly in provincial regions, simulated intelligence can assist with alleviating the effect of asset inadequacies in under-served or rustic regions. For instance, if a country region doesn't have simple admittance to radiologists, simulated intelligence can audit x-beam's and give indicative proposals. Computer-based intelligence symptomatic obligation help can help in different medication spaces and can help clinicians in under-resourced locales or areas. Once more, this innovation is to be utilized related to clinicians to assist with working on the exactness of findings and give better consideration.

Medical care is an exceptionally intricate world, and likewise, with any innovation, simulated intelligence in medical care is being investigated and refined. As simulated intelligence keeps developing, it will keep on supporting human suppliers in giving better and quicker consideration while likewise lessening costs. Moreover, this innovation will reach out to various spaces of medical services in the coming years, changing medical services all in all just as individual patient consideration. Therefore, all medical services associations should start understanding the advantages and capacities of artificial consciousness to remain on top of things and be ready for the capacities to come.

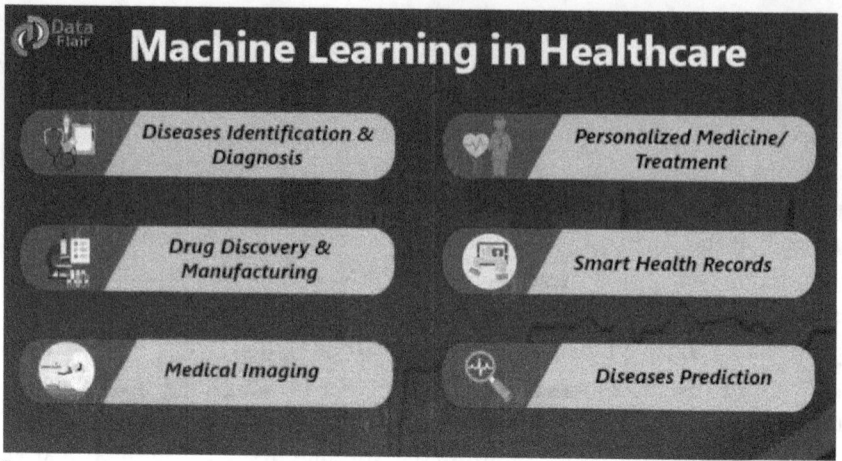

Source: data-flair.training

How Machine Learning improves data and informatics in healthcare

You should know that there are three key areas in which machine learning in health informatics impacts healthcare, and they include:

a) Recordkeeping

Machine Learning in well-being informatics can smooth out recordkeeping, including Electronic Health Records (EHRs). Utilizing simulated intelligence to develop EHR further, the board can work on quiet consideration, decrease medical care and authoritative expenses, and enhance tasks. For example, one model incorporates regular language preparing, which empowers doctors to catch and record clinical notes, killing manual cycles. AI calculations can likewise make EHR the executives' frameworks simpler to use for doctors by giving clinical choice help, robotizing picture investigation, and incorporating telehealth innovations.

b) Information Uprightness

Serious lacunas in health care enable the presence of data that facilitates Machine Learning analysis which leads to making base predictions, which can affect dynamism in the health care

28

environment. Because health information is initially planned for EHRs, it needs to be well gathered before ML calculations can productively utilize it. Healthcare informatics professionals are held accountable for keeping up with information honesty. Healthcare informatics professionals' activities incorporate get-together, dismantling, ordering, and pruning the data resources before final predictions.

c) Prescient Examination

The mix of machine learning, well-being informatics, and proactive investigation offers freedoms to develop medical care measures further, change clinical choice help apparatuses, and work on quiet results. The guarantee of AI's changing medical care lies in its capacity to use well-being informatics to foresee well-being results through prescient investigation, prompting more exact analysis and therapy and further developing doctor bits of knowledge for customized and accomplice therapies. AI can likewise offer extra benefit from a prescient investigation by interpreting information for chiefs to uncover measure holes and further develop medical services business tasks.

d) Resources

The accompanying assets can give a more prominent comprehension of the connection between AI and well-being informatics:

Public Place for Biotechnology Data, "machine learning and Electronic Wellbeing Records: A Change in outlook": This investigation examines the utilization of AI techniques to EHRs and its potential in further developing determination and treatment.

Forbes, "The 9 Greatest Innovation Patterns That Will Change Medication and Medical care in 2020": This article features the greatest and most extraordinary innovation patterns in medical services, including AI.

HIMSS, "Computerized reasoning in Well-being: Moral Contemplations for Exploration and Practice": However medical care innovation development keeps on changing medication, one should consider AI's moral ramifications.

Gov, Wellbeing IT Educational program Assets for Teachers: More than 10,000 instructors throughout the planet have utilized this asset to get to well-being IT educational program.

Wired, "From Determination to All-encompassing Patient Consideration, AI Is Changing Medical care": This article gives instances of how AI is changing medical services.

CHAPTER THREE

SPECIFIC APPLICATION OF MACHINE LEARNING AND ARTIFICIAL INTELLIGENCE IN HEALTHCARE

Specific ways of applying Machine Learning in Health Care

Machine Learning can decidedly affect patient consideration conveyance systems. For instance, it can help clinicians distinguish, analyze, and treat illness. Utilizing AI in medical services can likewise smooth out medical services undertakings and upgrade a medical procedure arranging, planning, and execution. The increment in the quantity of Machine Learning applications permits us to investigate the future where medical care suppliers will utilize information and examination to offer better types of assistance, further develop measures, and robotize errands. Before long, AI-based applications working with ongoing patient information will become ordinary, expanding the effectiveness of treatment alternatives, and driving the expense of medical services down. Here are some ways:

a) **Handling administrative tasks with Natural Language Processing**

The best weight doctors experience today is the association and execution of authoritative assignments. Via mechanizing them, medical services organizations could assist with taking care of the issue and permit doctors to do what they specialize in: invest more energy with patients. Here's an illustration of how Machine Learning and

Natural Language Processing can help. An enormous segment of authoritative undertakings includes assessing and refreshing electronic well-being records. Every clinic in the US uses a particularly bound together framework, and most facilities do too. Organizations could execute NLP-fueled apparatuses that would utilize calculations to distinguish and arrange words and expressions to permit doctors to direct notes straightforwardly to the framework during patient visits. A short time later, specialists could survey the graphs and synopses arranged by the device instead of perusing every one of the notes and test results to comprehend a patient's general strength. The device would assist doctors with investing energy in keeping up with the patient records, thus permitting them to convey better medical care administrations to patients.

Source: canopylab.com

b) Resources

Medical services suppliers currently exploit computerized arrangements based on top of Machine Learning models that utilize irregularity identification calculations to anticipate occasions like strokes, coronary failures, sepsis, and other genuine entanglements. These instruments pull information from verifiable patient records, daily assessments, and continuous estimation of imperative signs, such as the pulse or circulatory strain. As a result, the devices can make staff aware of inevitable patient dangers and permit them to make preventive moves.

For instance, in El Camino Clinic, scientists consolidated electronic well-being records, nurture call information, and bed alert information to foster a device for foreseeing patient falls. This device cautions staff when the patient is in great danger of falling with the goal that they can quickly make a move. The execution of the device assisted with lessening falls by 39%.

c) Speeding up clinical examination

To stay aware of the patterns, specific spaces of clinical examination, doctors and researchers need to peruse and handle a mind-boggling amount of data. In addition, researchers distribute a great many examination papers each year, and staying aware of the latest exploration is regularly difficult. By utilizing a Natural language processing (NLP) apparatus to parse writing, clinical specialists could get important experiences without perusing the entirety of the articles all alone.

For instance, scientists are from Ireland and the US worked together on an investigation on antagonistic medication occasions utilizing text mining, neural organizations, and prescient examination to break down immense information bases of clinical writing just as web-based media presents on to see how individuals are remarking on incidental drug effects. This strategy permitted them to investigate more than 300,000 articles from clinical diaries and 1.6 million remarks via online media. Furthermore, the group utilized some convenient information perception instruments to show the connection between medications and incidental effects. As you can envision, physically completing an errand of this greatness would take the group years, if not many years.

d) Determination and infection distinguishing proof

The main uses of Machine Learning calculations in medical services are identified with the ID and finding of sicknesses that are thought about difficult to analyze. This can incorporate malignant growths that are hard to distinguish during their underlying phases of hereditary illnesses. An illustration of such an application is IBM Watson Genomics, which coordinates intellectual registering with genome-based tumor sequencing to help doctors make a quick analysis. There are additional instruments that exploit artificial reasoning to foster remedial medicines in regions like oncology. The thought here is fostering an industrially suitable approach to analyze and give treatment in clinical conditions via robotizing the interaction however much as could be expected.

e) Learning and assembling drugs

The beginning phase drug disclosure measure is another region that can profit a ton from AI. It's now overwhelmed by Research and development innovations, such as accuracy medication and cutting-edge sequencing, which is utilized to discover elective ways to treat multifactorial infections. In addition, since AI methods depend on unaided picking up (distinguishing designs in information without giving any forecasts), they can be exceptionally valuable for finding new medications and customizing drug mixes for explicit cases.

f) Clinical imaging diagnostics

Machine Learning and Deep learning brought us advancement innovation called PC vision. Numerous tech organizations all throughout the planet are presently bustling, creating devices that give diagnostics to picture examinations for doctors. As AI calculations become more inescapable and fill in their ability, we will see an expanding number of information sources from different clinical symbolism.

For instance, quite possibly, the main application we have seen so far is the investigation of skin pictures that expect to distinguish skin malignant growth. In a few investigations, such apparatuses were demonstrated to be more exact than doctors. For example, they arrived at 87-95% exactness while dermatologists have a 65% to 85% precision rate in distinguishing melanomas.

g) Personalization

Medicines are best when they're joined with singular well-being factors. That is the reason AI, and its prescient examination segment can assume a particularly immense part in customized medicines.

Presently, doctors can look over a restricted arrangement of findings or gauge the danger to their patients based on their indicative history and the accessible hereditary data. Later, AI instruments will want to use patient clinical history to create numerous treatment choices. We are likewise going to see more gadgets and biosensors with cutting-edge well-being estimation capacities show up available. They will consider information to open for ML-based advancements.

h) Health records

Keeping up with and refreshing health records is a tedious and costly cycle. The facts confirm that innovation plays had a fundamental impact in working with the information section measure. In any case, most cycles set aside a ton of effort to finish since they should be done physically. This is the place where AI comes in. It vows to save time, cash, and exertion. For instance, record arrangement strategies that utilization vector machines and AI-based OCR acknowledgment are currently showing up available. An illustration of that is the Google Cloud Vision Programming interface or Matlab AI-based penmanship acknowledgment innovation. In addition, different organizations are currently fostering the up-and-coming age of shrewd well-being records that will fuse ML apparatuses from the beginning to help in the clinical treatment ideas and conclusion.

i) Flare-up expectation

Machine Language-based apparatuses are currently additionally being utilized to screen and anticipate flare-ups throughout the planet. Did you realize that BlueDot, a specific apparatus for observing expected flare-ups, anticipated the spread of the Covid-19 before it has been authoritatively declared? Researchers can survey a gigantic measure of information today gathered from continuous online media that takes care of satellites, site data, and institutional records. Organizations can assist with sorting out this data and foresee anything from jungle fever flare-ups to serious irresistible infections. Anticipating such episodes is particularly important for Underdeveloped nations that need the clinical framework and instructive frameworks.

You never neglect to focus on how Computerized reasoning and AI will affect the two doctors and emergency clinics sooner rather than later. They will assume a basic part in clinical choice help, illness recognizable proof, and fitting treatment intends to guarantee the ideal results. We are additionally going to utilize AI-based apparatuses to offer diverse treatment choices, customized medicines, and working on the general productivity of emergency clinics and medical care frameworks while decreasing the expense of care.

CHAPTER FOUR

DOWNSIDES TO MACHINE LEARNING APPLICATION IN HEALTHCARE

In the flicker of an eye, notices of artificial brainpower have gotten universal in the medical services industry. From deep learning measurements that can scan through CT checks faster and more efficiently than people to natural language processing (NLP) that can sift through unencumbered data in electronic health records (EHRs), the applications for computer-based intelligence in medical services appear to be unending. Be that as it may, like any innovation at the pinnacle of its publicity bend, Machine Learning, a subset of artificial reasoning, faces analysis from its doubters close by energy from fanatic crusaders. Despite its ability to open new pieces of information and smooth out how providers and patients' partner with clinical benefits data, it may bring not superfluous risks of assurance issues, ethics concerns, and clinical mistakes. Interest in artificial intelligence for clinical benefits has grown colossally, recalling work for diagnosing diabetic retinopathy, distinguishing lymph center point metastases from chest pathology, mental lopsidedness, and tremendous degree phenotyping from observational data. Despite these advances, the prompt use of simulated intelligence to clinical benefits remains loaded with ensnarement. Huge quantities of these troubles come from the apparent target in clinical consideration to make tweaked assumptions using data created and directed through the clinical system, where data arrangement's principal job is to help care instead of work with resulting examination.

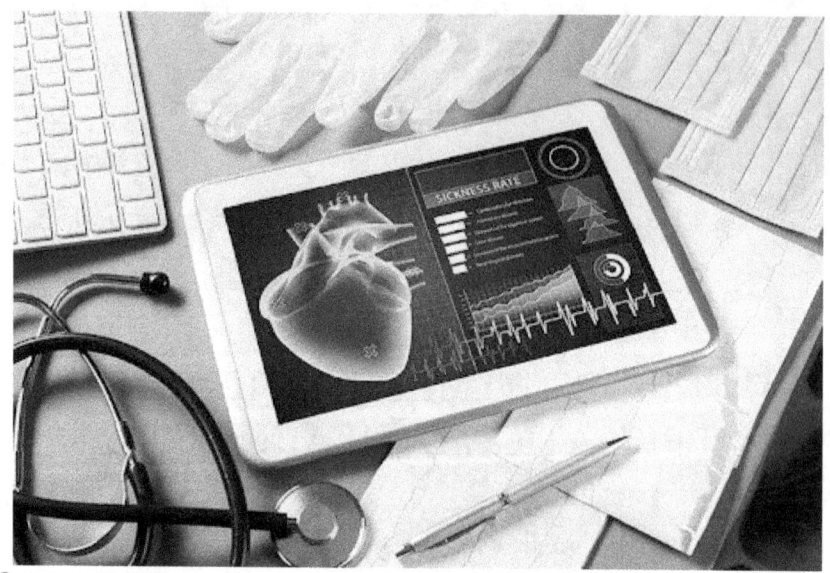

Source: allerin.com

The disadvantages associated with machine learning in healthcare thus:

a) Injuries and error

The clearest peril is that artificial reasoning structures will now and again not be correct. That patient injury or other clinical benefits issues may result. If an artificial brainpower system proposes some unsuitable drug for a patient, falls to see a tumor on a radiological yield, or apportions a crisis center bed to one patient over another because it expected wrongly which patient would benefit more, the patient could be hurt. Various injuries happen because of clinical bungle in today's clinical benefits structure, even without the consideration of reproduced knowledge. PC-based insight bumbles are potentially one of a kind for somewhere near two reasons. First, in any case, patients and providers may react particularly to wounds coming about due to programming than from a human mix-up. Second, whenever reenacted insight systems become certain, a fundamental issue in one artificial consciousness structure might achieve wounds to a considerable number of patients—rather than the set number of patients hurt by any single provider's screw-up.

b) Data openness

Planning artificial consciousness systems requires a great deal of data from sources, for instance, electronic revenue records, pharmacy records, assurance claims records, or client-created information like well-being trackers or purchasing history. However, prosperity data are routinely perilous. Data are routinely separated across a wide scope of systems. Undoubtedly, even next to the combination just referred to, patients consistently see different providers and switch protection offices, provoking data split into various systems and various associations. This brokenness fabricates the risk of bungle, lessens the breadth of datasets, and grows the expense of get-together data— which also limits such components that can encourage effective clinical benefits of PC-based knowledge.

c) Insurance concerns

Another plan of perils arises around privacy. The need for colossal datasets makes inspirations for fashioners to assemble such data from various patients. A couple of patients may be concerned that this arrangement may manhandle their security, and cases have been reported ward on data splitting between colossal prosperity systems and recreated knowledge engineers. The reproduced insight could entrap insurance in another way: artificial reasoning can expect private information about patients despite how the estimation never got that information. (Without a doubt, this is consistently the goal of clinical benefits artificial consciousness.) For instance, an artificial brainpower system might recognize that an individual has Parkinson's sickness subject to the shaking of a PC mouse, whether the individual had never uncovered that information to some other individual (or didn't have even the remotest clue). Patients ought to truly think about this as an encroachment of their security, especially if the artificial knowledge structure's acceptance was available to pariahs, such as banks or life inclusion associations.

d) Inclination and dissimilarity

Perils inferring tendency and lopsidedness in clinical consideration of computerized reasoning. Recreated knowledge structures acquire from

39

the resource on which they are ready, and they can join inclinations from those data. For instance, if the data open for recreated insight are principally gathered in academic clinical centers, the ensuing artificial knowledge systems will contemplate and as such will treat less reasonably—patients from masses that don't normally visit educational, clinical core interests.

Whether or not artificial consciousness structures are acquired from accurate, specialist data, there can regardless be issues if that information reflects stowed away tendencies and irregular characteristics in the prosperity system. For example, less treatment for torture than white patients;8 an artificial knowledge system acquiring from prosperity structure records might sort out some way to prescribe lower doses of painkillers to African American patients regardless of how that decision reflects crucial tendency, not regular reality. Resource assignment artificial knowledge systems could, in like manner, fuel unevenness by giving out fewer resources for patients considered less charming or less useful by prosperity structures for a variety of risky reasons.

e) Proficient realignment

Longer-term chances remember shifts for the clinical calling. Some clinical qualities, like radiology, will move impressively as a lot of their work becomes automatable. A couple of analysts are concerned that the all-over usage of PC based insight will achieve decreased human data and cutoff as time goes on, with the ultimate objective that providers lose the ability to get and address artificial knowledge bumbles and further to encourage clinical data.

CHAPTER FIVE

FUTURE OF MACHINE LEARNING APPLICATION IN HEALTHCARE

With Machine Learning, medical care specialist co-ops can settle on better choices on drug determinations and therapy alternatives, which lead to a general improvement of medical care administrations. Already, it was trying for medical care experts to gather and examine the enormous volume of information for successful expectations and therapies since there were no advances or instruments accessible. Presently with AI, it's been moderately simple, as large information advancements, for example, Hadoop, are experienced enough for wide-scale reception. For sure, 54% of affiliations are using or considering Hadoop as an enormous data-getting-ready gadget to get huge pieces of information on clinical benefits, according to the Ventana Investigation Study. Moreover, 94% of Hadoop customers out of existing customers examine voluminous data, which they acknowledge was unreasonable already. Artificial intelligence estimations can be valuable in giving vital estimations, persistent data, and advanced examination to the extent of the patient's infection, lab test results, beat, family lineage, clinical starter data, etc., to trained professionals.

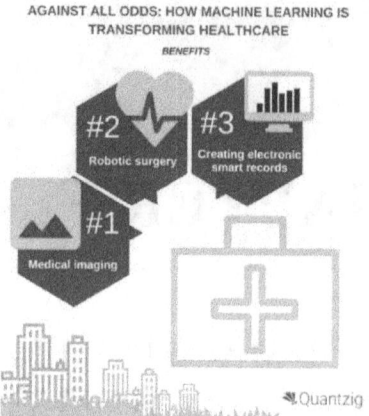

Here are some prominent future successes to be recorded in the health care sector deploying Machine Learning:

Disease Prediction

The advanced way to deal with medical services is to forestall the infection with early mediation instead of going for therapy after analysis. Customarily, doctors or specialists utilize a dangerous minicomputer to survey the chance of infection improvement. These minicomputers utilize principal data like socioeconomics, ailments, life schedules, and more to figure the likelihood of fostering a specific sickness. Such computations are finished utilizing condition-based numerical techniques and devices. The test here is the low precision rate with a comparative condition-based methodology. For instance, the Framingham Study can anticipate the hospitalization with just fifty-six Percent of exactitude for a drawn-out cardiovascular illness.

Professionals in this discipline deal with the philosophies to recognize, create, and tweak ML projections and designs that ensure precise forecasts. To foster a solid and more precise AI model, we can utilize information gathered from considers did, patient socioeconomics, clinical well-being records, and different sources. The distinction between customary and Machine Learning approaches for illness expectation is the number of ward factors to consider. In a conventional methodology, they consider not many factors that you can depend on your figure like age, weight, stature, sex, and that's just the beginning (because of computational constraint). Then again, AI being prepared on processing gadgets can think about an enormous number of factors, which brings about a superior precision of medical services data. Consumer wearables and other clinical gadgets give an unmistakable thought that the eventual fate of computerized reasoning in medical care is brilliant. The innovation is being utilized to screen and recognize dangerous scenes in the beginning phase of coronary illness, permitting specialists and different guardians to more likely screen and distinguish dangerous scenes at a prior, more treatable stage.

Drug Discovery

Medicinal revelation and headway are over the top and drawn-out work. Consistently, another prescription landslide success needs more than ten years to get into a market and costs around 2.6 billion dollars, as demonstrated by the Tufts Spot for the Examination of Medicine Improvement. A prescription report drive is highlighted, finding a compound that reacts with the assigned particles of the body, making a disease fix. However, there is an exceptional possibility that the middle or supporting drug compound reacts to non-assigned particles in the body unfairly, which may cause compromising and unsafe accidental impacts.

As medication associations can't predict a potential prescription compound effect on assigned and non-assigned iotas using standard computational progressions, the chances of medicine dissatisfaction are higher in clinical primers. Moreover, the present circumstance makes drug revelation costly and drawn-out collaboration. Therefore, better-farsighted procedures using artificial intelligence can save a lot of resources for the present circumstance.

AI-based system (pondering the immense volume of clinical data for supported and bombarded drugs) to recognize a poisonous compound that may cause accidental impacts can save various resources before going into clinical primers. Around 90% of drugs can't bear the primer cycle. By automating the compound particle's reaction measures using simulated intelligence, medications can also foster the prescription divulgence and progression cycle and reduce the chance to promote. According to an assessment employing Carnegie Mellon, mechanizing the drug exposure cycle can diminish costs by 70%.

Enhanced Patient Experience

Presenting simulated intelligence and ML innovations in medical services aims to upgrade the supplier and patient experience. Without artificial intelligence incorporated in innovation, we see medical care groups embrace various items to finish work processes, including virtual well-being enablement, group coordinated effort, local area cooperation, and virtual adjusting. This is neither effective nor useful. A solitary, incorporated arrangement presents efficiencies, diminishes burnout, further develops fulfillment, and results, and decreases the general expense of conveying care.

Simulated intelligence remote helper bots help in the beginning stages of virtual well-being visits. Their job in first-stage patient virtual visits begins in the virtual lounge area. The best advantage of the computer-based intelligence bot is its impact on diminishing the measure of time spent standing by to see a doctor. At the same time, it consequently starts numerous authoritative errands, like patient admission and emergency, e-assent frames, and giving patients instructive substance that would be the same as an in-person experience. The help, this way from this simulated intelligence innovation, lessens the responsibility for the regulatory staff, who would then be able to take on the expanded consideration limit concerning the doctors in the facility.

Effective Health Delivery

Most fundamentally, in future occasions to come, ML models can assist doctors with diagnosing patients, particularly in cases including moderately uncommon illnesses or when results are difficult to foresee. For instance, in a new clinical investigation, a few AI models were utilized to dissect information from electronic well-being records to anticipate cardiovascular breakdowns. The results showed that these ML frameworks anticipated results well. Additionally, ML can decide the best drug measurement, lessening medical care costs for the patients and suppliers. ML can be utilized in deciding measurement and deciding the best drug for the patient. Hereditary varieties among various races, nationalities and unique individuals affect the viability of specific medications and individuals' reactions to these medications, like HIV meds. Again, progressed ML calculations and models are created. They would have the option to quickly perceive these distinctions and arrive at exact and dependable resolutions.

Employing Machine Learning to Better Recognize the People Behavior

Clinics associated with one another and to one another's patients. Alongside prescient consideration, there's another development in such consideration. In 2030, a clinic is as of now not an enormous solitary design that treats a wide scope of illnesses; all things being equal, it centers care around the sick and exceptionally complex methodology, while less critical cases are observed and treated through more modest centers and spokes like retail facilities, same-day medical procedure places, expert therapy centers, and surprisingly individuals' homes. Abound together advanced framework interfaces these spots, which further shows the positive commitment of incredible computer-based intelligence in the medical services future. Brought together, war rooms dissect clinical and geographic information continuously to monitor the organic market across the organization. This organization can ease bottlenecks in the framework and assurance that patients and the medical care workforce are coordinated to where they can best be really focused on or where they are required and utilize artificial intelligence to perceive individuals at risk of decay. The area is, as of now, not the magic that binds this organization. The encounters of the people it serves will decide the third significant contrast in 2030.

Resistance To Antibiotics: Management of The Risks

Information from electronic well-being records can help with the location of disease designs and the ID of individuals in danger before they foster indications. Utilizing AI and artificial intelligence to drive these examinations can improve precision and give medical care suppliers quicker, more exact cautions.

Promoting The Use of Immunotherapy in The Treatment of Cancer

Immunotherapy is quite possibly the most encouraging way to deal with malignancy treatment. Patients might have the option to overcome troublesome tumors by using the body's insusceptible framework to attack them. Nonetheless, existing immunotherapy options just advantage a small level of patients; oncologists do not have an exact and precise methodology for determining which patients might profit from this treatment. AI calculations and their capacity to blend extremely complex data might open the entryways for the eventual fate of artificial brainpower in medical care.

CONCLUSION

Machine Learning, no doubt, has immensely reduced the so many errors associated with human inputs and activities in the medical service delivery sector of the larger society. This aspect of artificial intelligence has equally embraced an enlarged portion of health data analysis, thereby boosting its abilities to make predictions about disease outbreaks or even the probability of health disasters. Machine learning has also aided in the aspect of drug discovery which is patently relevant to diseases and ailments par time, which helps boost the confidence and experience of patients. More interestingly, machine learning holds very great potentials for the future of the health care sector upon its full exploration and further advancement. You should note that the era of human errors was tolerated when individuals' well-being is at stake is no longer tolerated. Hence, there is a need for more precision and accuracy in treatment, especially through a preventive approach.

REFERENCE

1. *What is artificial intelligence (ai)?* Oracle. (n.d.). https://www.oracle.com/artificial-intelligence/what-is-ai/.
2. *How machine learning works.* Algorithmia Blog. (2021, May 17). https://algorithmia.com/blog/how-machine-learning-works.
3. Kumar, S. (2019, December 12). *Advantages and disadvantages of artificial intelligence.* Medium. https://towardsdatascience.com/advantages-and-disadvantages-of-artificial-intelligence-182a5ef6588c.
4. Bresnick, J. (2019, December 18). *5 steps for planning a Healthcare artificial intelligence project.* HealthITAnalytics. https://healthitanalytics.com/features/5-steps-for-planning-a-healthcare-artificial-intelligence-project.

5. B. D., By, Dickson, B., -, Ben DicksonBen is a software engineer and the founder of TechTalks. He writes about technology, Ben is a software engineer and the founder of TechTalks. He writes about technology, & var block_td_uid_6_611a70de5cebd = new tdBlock();block_td_uid_6_611a70de5cebd.id = "td_uid_6_611a70de5cebd";block_td_uid_6_611a70de5ceb d.atts = '{"limit":6. (2021, January 4). *What is semi-supervised machine learning?* TechTalks. https://bdtechtalks.com/2021/01/04/semi-supervised-machine-learning/.
6. *Top 10 applications of Machine Learning: Daily life applications.* Edureka. (2021, July 29). https://www.edureka.co/blog/machine-learning-applications/.

7. W. N. P. (2020, May 6). *Risks and remedies for artificial intelligence in health care.* Brookings. https://www.brookings.edu/research/risks-and-remedies-for-artificial-intelligence-in-health-

care/#:~:text=While%20AI%20offers%20a%20number,health%2
Dcare%20problems%20may%20result.&text=AI%20errors%20are
%20potentially%20different%20for%20at%20least%20two%20rea
sons.

8. Davenport, T., & Kalakota, R. (2019). The potential for artificial intelligence in healthcare. *Future Healthcare Journal*, *6*(2), 94–98. https://doi.org/10.7861/futurehosp.6-2-94

9. *Machine learning in HEALTHCARE: EXAMPLES, tips & Resources: UIC Online*. UIC Online Health Informatics. (2021, July 22). https://healthinformatics.uic.edu/blog/machine-learning-in-healthcare/.

10. Lorberfeld, A. (2019, April 25). *Machine learning algorithms in layman's terms, part 1*. Medium. https://towardsdatascience.com/machine-learning-algorithms-in-laymans-terms-part-1-d0368d769a7b.

11. Menon, K. (2021, June 24). *Types of machine learning: Simplilearn*. Simplilearn.com. https://www.simplilearn.com/tutorials/machine-learning-tutorial/types-of-machine-learning.

12. *Pros and cons of artificial intelligence - a threat or a blessing?* DataFlair. (2018, January 16). https://data-flair.training/blogs/artificial-intelligence-advantages-disadvantages/.

13. *Machine learning in healthcare - unlocking the full potential!* DataFlair. (2021, March 8). https://data-flair.training/blogs/machine-learning-in-healthcare/.

14. *Top benefits of machine learning in the healthcare industry: Quantzig*. Business Wire. (2018, August 2). https://www.businesswire.com/news/home/2018080200541 0/en/Top-Benefits-of-Machine-Learning-in-the-Healthcare-Industry-Quantzig.

15. Joshinav. (2018, April 16). *Top 5 applications of deep learning in healthcare: Artificial intelligence*. Application development. https://www.allerin.com/blog/top-5-applications-of-deep-learning-in-healthcare.

www.ingramcontent.com/pod-product-compliance
Lightning Source LLC
Chambersburg PA
CBHW070948200526
45161CB00001BA/28